MW00901173

Bless This Mess

A Picture Story
of Healthcare in America

Stephen K. Klasko, MD, MBA

Illustrated by Chrissie Bonner
With Michael Hoad

Copyright © 2018 Stephen K. Klasko, MD, MBA

All rights reserved. No part of this book may be reproduced, stored, or transmitted by any means—whether auditory, graphic, mechanical, or electronic—without written permission of the author, except in the case of brief excerpts used in critical articles and reviews. Unauthorized reproduction of any part of this work is illegal and is punishable by law.

Jefferson University Press

Cover design by Klaus Herdocia

Interior Art Credit: Chrissie Bonner

ISBN: 9781-4834-7962-0 (sc)
ISBN: 978-1-4834-7961-3 (hc)
ISBN: 978-1-4834-7960-6 (e)

Library of Congress Control Number: 2018900267

Because of the dynamic nature of the Internet, any web addresses or links contained in this book may have changed since publication and may no longer be valid. The views expressed in this work are solely those of the author and do not necessarily reflect the views of the publisher, and the publisher hereby disclaims any responsibility for them.

Lulu Publishing Services rev. date: 03/21/2018

Bless This Mess

5.
A Hitchhiker's Guide to the BEST of the Heathcare Galaxy

6.
The Prime Disruption ...or... Captains Say the Darndest Things

7.
YOU Design the Future

Contents

Chapter 1 The USA on the Planet Earth in the Galaxy Called The Milky Way 1

1.1 Why They Don't Sell Milky Ways in Whole Foods in America ... and Why American Healthcare Will Also Get Bought By Amazon If We Don't Join the Consumer Revolution 1

Chapter 2 Whose Fault Is It? How High can You Count? 11

2.1 From Marcus Welby to House 12

2.2 You're a Lot Safer Flying With Me…Than Being Operated On By Me 17

2.3 The "C" Word: Forbidden Around the Galaxy (except for the USA) 19

2.4 It's Hard to Get Someone To Do Something When Their Salary Depends Upon Them Not Doing It 24

3.1 Healthcare's Customer Service Is a Real Turkey 33

3.2. Hey BUB, I Have No Idea What You Charged Me For 44

3.3: Death, Dialysis and Decisions That Have To Be Made (even in Alaska) 48

Chapter 3 Yes, You Should Have Paid Attention in Ninth Grade Geometry Class 51

Chapter 4 A Hitchhiker's Guide to the Best of the Healthcare Galaxy 59

TARVOS 60

VENDRIZI 3 61

ROYA VOSA 62

YOVIDO 63

AMISIS 64

TERONDA PRIME 65

CRIUS 66

TANDRA 67

OCTAVIA 68

NYOTA 5 69

Chapter 5 The Prime Disruption 71

Conclusion 77

"Do ... or do Not ... There is no Try." Yoda

Girls and Boys, Ladies and Gentlemen
This is a story of hope
for the country you will inherit
with a healthcare system
that is a mess.

We created this book because we believe *you* can make a difference, no matter how young or old, no matter your gender, no matter where you come from. Most importantly, you can make a difference whether you have a traditional "leadership" job, or whether you simply believe in transformation. There are no limits to what we can do – when we decide there are no limits.

Larry Page, co-founder of Google, once said, "The main thing that has caused companies to fail is that they have missed the future."

We couldn't agree more.

Who are we?

I'm Steve Klasko. I'm a doctor who came to believe in the power of transformation. I run a university and a large health system, and I tell everyone I meet that we can transform healthcare in America. We know how. But we need to "Do." As Yoda said, "Not try."

Chrissie Bonner is an artist and a generative scribe. She draws pictures to help groups understand their stories. Chrissie worked with a former journalist named Michael Hoad to put the book together.

How we met: One day in Philadelphia Steve gave his talk, "The hitchhiker's guide to the healthcare galaxy." Chrissie was the "scribe," who drew the talk while he spoke. This technique helps people visualize ideas, and is often used in team building. In a stroke of genius, Chrissie drew a picture of the book, "Bless This Mess," leading to our collaboration.

Here is Chrissie's sketch from that original meeting. It formed the basis for this book.

The most important author, however, is you. You are the reader who can take your ideas and put them on these pages.

Draw what you believe. Write. Speak. Act. Most importantly, demand. Because the simple fact is that you accept from healthcare what you would not tolerate in any other venue. If you had tolerated in retail what you tolerate in healthcare, then Macys, Sears, Penneys and other retailers would still be ruling the world.

Amazon did not change the retail world … it was the fact that most retailers did not understand the changing needs of consumers.

Netflix did not change the entertainment world … it was the fact that Blockbuster never realized that the product was the entertainment, not the store.

Uber and Air BnB did not change ride leasing and the hotel industry … it's just that they gave what consumers needed, and in some cases needed to explain to consumers how things could be better.

This is an optimistic story about how YOU can make a difference. How you can demand what you now get in every other sector of the economy. And how, because of you, we - the purveyors of your healthcare - finally got it.

We're going to show you why YOU are the most important author in writing the history of the future of healthcare in America.

And how because of you, a wonderful thing happened. We were inducted into the intergalactic council of high functioning, consumer-centric, outcomes-driven planetary healthcare systems. Another wonderful thing happened … we finally figured out the cost, access, patient experience, and quality curve. More on that later.

The book will also give you a perspective of how extra-planetary beings look at us. And it will give you a perspective of how other planets in our galaxy were able to look outside the box or whatever geometric figure occurred on those planets.

We will also start by telling you how they graded our healthcare system in 2018.

Spoiler alert: You would not have wanted to bring that report card home to your parents when you were in high school.

Then we're going to imagine how, together, we can apply those solutions. We're going to imagine what a great future would look like in 2035. And we're going to make that future happen now.

Acknowledgements

With many thanks to David Klasko who took all my worst jokes out of this book. To Jill and Lynne, for making sure that anything I write applies to ALL people. To Colleen who reminds me daily why we fight for a better future. And to Chrissie Bonner and Michael Hoad for making it happen.

Stephen Klasko

Prologue: A Shocking Decision by the Inter-galactic Health Council

We begin with shocking news: It's 2035 and the Intergalactic Health Council has invited the United States of America to join the Council as an exemplar of great healthcare for all its people.

Meeting holographically across billions of planets, Council members were themselves shocked as representatives of the USA gave their report.

On the beautiful Planet Earth, they began. In the Galaxy called the Milky Way, a race called humans had developed healthcare delivery that is equitable, personal, driven by community interaction and technologically advanced with a human touch. Most importantly, health is now considered a team sport involving the person and their entire team of partner-providers.

"That's impossible," a delegate from a trillion light years away intoned across its non-verbal communication device.

"In its year 2018, the USA was known for hospital-centered, fee for service rescue medicine, with evidence gathered by research by individual institutions who often did not communicate with each other or share data, with huge disparites between African Americans and Caucasian Americans so great as to create gaps of up to 20 human years in lifespan. This was especially surprising given that earth was inhabited by a relatively simple set of sentient beings with only two genders (the average among the council was planets with seven genders.[1]) Even on its own planet, the USA was known for spending the most with the least return."

"It is IMPOSSIBLE that the USA would be considered an example of great healthcare in so short a time. What on earth changed?"

This book, *Bless This Mess*, is the USA's report and attempt to explain to the council that this is not a hoax but rather an orchestrated vision and implementation that started in 2018.

It is a primer of how to transform a healthcare delivery system for an entire nation on one planet.

We begin with a look at how other species on other planets saw the country called the USA, on the planet called Earth, in 2018.

Then we tour how other planets transformed healthcare.

Then we'll come back to how change happened in the USA, and how every one of you was part of that revolution.

The USA on the Planet Earth in the Milky Way

---

[1] Aliens watching Earth has been documented often (see *A Wrinkle in Time*, by Madeleine L'Engle in 1962).

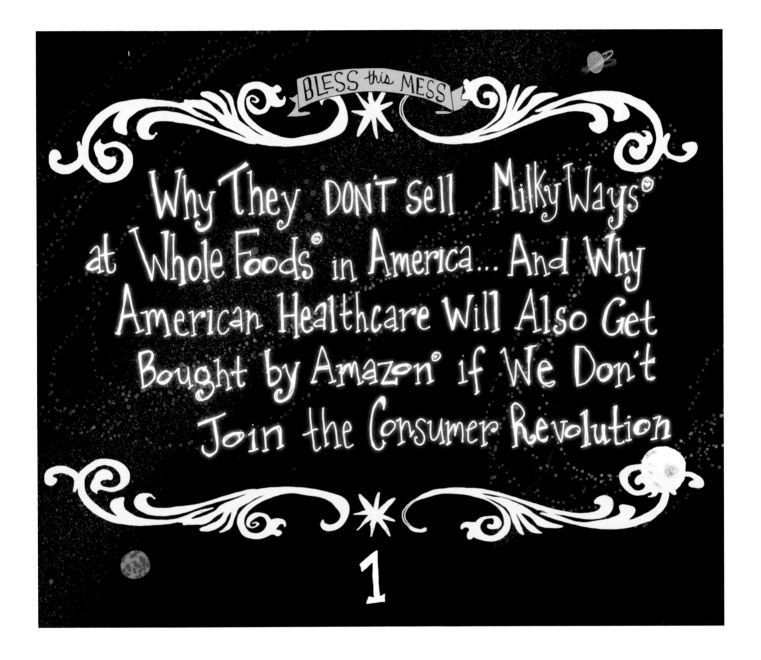

BLESS this MESS

Why They DON'T Sell MilkyWays at Whole Foods in America... And Why American Healthcare Will Also Get Bought by Amazon if We Don't Join the Consumer Revolution

1

From deep space, aliens see a beautiful country named the United States. (They see other beautiful countries too, but let's start with this one.)

The United States is on the Planet Earth, which is also wonderful and beautiful.

Planet Earth is in a galaxy called Milky Way. (We think the Milky Way is beautiful too, but we're not sure since Uber does not yet travel to its other 200-billion stars.)

The aliens got the idea that the Milky Way was named after a candy bar in the USA. Yes, that candy bar, the one with 9 grams of fat, 31 grams of sugar and 75 grams of sodium.

It was an easy mistake to make because what the aliens saw was scary. They saw lots of people walking on the street eating candy bars and other even more unhealthy foods.

Can you name these three sources of nutrition in the USA (planet earth)?

If you can name these sources of nutrition, then you know that the beautiful country USA on the beautiful planet Earth has more overweight people than any place in the galaxy because the humans in that country eat more fat, sugar and sodium than any place else in the Milky Way (the galaxy), partly because of the Milky Way (the candy).

One reason healthcare in America is a mess is because too much of that yummy food can cause an evil monster to invade your pancreas and wreak havoc with your body. In some cases, the leading cause of diabetes is often preventable with good nutrition. Unfortunately, the USA on the planet Earth is the only place in the galaxy where nutrition is undervalued and underpaid for, but there is gobs of money to take care of you after you've eaten too much of the wrong things.

For the people who live in the USA, many aspects of life have improved thanks to technology. People once went to bookstores to buy books, music stores to buy their tunes, and movie stores for DVDs. Now they can buy all of those from their phone. Instantly. No walking.

Pretty soon we may become too lazy to even turn on our TV and there will be a chip in our brain playing movies.

On the day after Thanksgiving[2], the people of the USA all used to get in their cars and wait hours to park and fight over cool toys to celebrate the holidays.

---

[2] Thanksgiving. The actual meaning of Thanksgiving is controversial but most Americans celebrate by killing large flightless birds and consuming way too much food.

Now they can do all their shopping while binge watching their favorite TV show (and in some cases binge eating some of those fat, sugary, salty goodies that invite diabetes.)

By the way, if you like Thanksgiving, we'll get back to it later in the book.

Technology has made the United States the envy of the galaxy by providing easy access to gadgets galore, online shopping, instant delivery, and endless TV programming, often at lower cost than in the past. But it hasn't succeeded in making one thing easier and more affordable to access. Can you guess the one area in the United States where technology has not improved the experience for the consumer? If you can't, please spend the net five minutes staring at the cover of this book. Concentrate on two words: mess and healthcare!

Healthcare is one part of the United States which has not become friendly to people at all. The fact that technology hasn't improved healthcare access and affordability is curious, because it's one thing that every human in the country needs.

In fact, it is very perplexing to the other inhabitants of the galaxy. For example, when people in the USA need to go to the hospital,[3] they get very funny bills which make no sense.

---

[3] Hospital. If you are new to this language, do not confuse "hospital" and "hospitality." They are very different.

And people don't mind because in most cases they actually don't pay the bills … other people do. They are called insurers. Insurers send funny and hard to understand bills also (that the people <u>do</u> have to pay).

Hospitals always talk about keeping people healthy but they often sponsor state fairs where the only source of nutrition is corn dogs, funnel cakes and milky ways (the fat, sodium and sugar ones, not the galaxy).

In the 21st century, more and more sick people from the United States of America didn't need to go to the big expensive hospitals because they could be cured in less expensive places, so the big expensive hospitals decreased their beds, right? No, they built new beds. Why? Because they could.

Oh, and even though technology now lets you get your books, music, movies, and shopping delivered when and where you need it, things still don't work like that in health care.

In most places in the United States, you still have to go to the "store" (aka, hospital) to get care, doctors still get paid to treat sick people instead of keeping them well, and while you can shop from home on the day after Thanksgiving, even if you have a stomach ache (from eating too much turkey), you cannot get that stomach ache treated from home.[4]

---

[4] In fact you probably will have to call a doctor's office and listen to 11 options before you can get an appointment next week, a process which usually results in a headache concomitant with your stomach ache.

So how did this happen?

Let's start with three teasers. Three things that don't exist on any of the other planets we will talk about. Planets that view healthcare as a right and not a privilege.

Oh, and each of those planets provide better care at a lower cost through the collaborative interaction of sentient beings.

The three teasers:

1 - OPM
2 - Geometry
3 - Fakish News

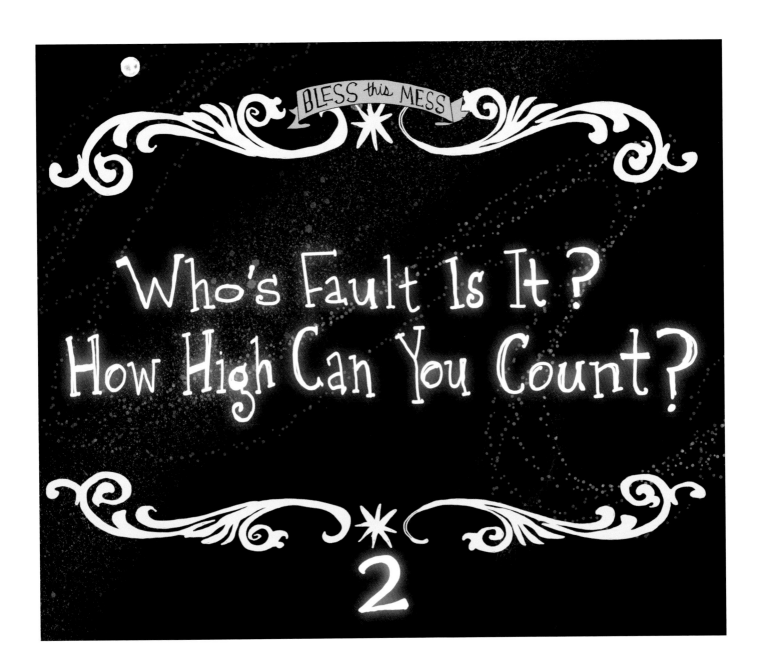

2.1 From Marcus Welby to House

This man is what people in the United States used to picture when someone said "doctor." His name is Dr. Welby. Notice his black bag. He makes house calls, stops to deliver puppies, sees patients in his primary care office and then goes to a hospital to perform complex neuro-surgery. This species of doctor is extinct.

This is today's doctor. His name is Dr. House. He is very smart, but he is not nice. He does not communicate well. He does not seem very empathetic. He is great at memorizing stuff.

How did we go from Dr. Welby to Dr. House in the United States healthcare system?

By choosing doctors based on science GPA, MCATs[5] and organic chemistry grades and ignoring empathy, communication skills and creativity. These doctors are very good at memorizing science and organic chemistry formulas and answering multiple choice questions. Unfortunately, patients are not multiple choice tests, so patients had to count on other people[6] to translate what their doctors are telling them.

You are probably thinking, "why don't they change the criteria so doctors can be more like the extinct Dr. Welby? Good question! It's partly because of this funny publication that ranks medical colleges and that every parent and every hoping-to-be medical student reads.

---

[5] For those of you who have never visited the United States on the planet Earth, the med-cat is not a living thing like other cats. It is not cute, soft or furry, in fact it is not even a very good indicator of who will be an empathetic, communicative, and creative healer.

[6] Other people as in nurses, who are more often selected and trained for holistic, empathetic DNA.

Those yearly rankings are based on … you guessed it … science GPA, MCATs, and organic chemistry grades. So, Dr. House graduated from one of those Top 10 medical schools. Are you starting to see the pattern?

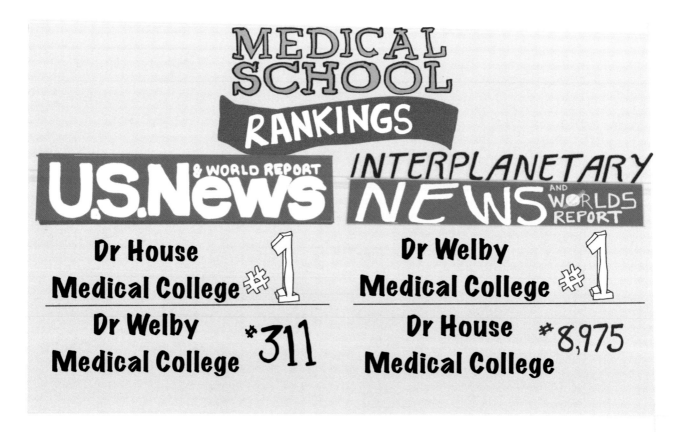

Other planets use a very different dashboard. It is called the *Interplanetary Report and Ranking* of medical schools. It is based on cool and important factors like emotional intelligence, cultural competence, creativity, communication skills, self-awareness, and empathy.

As we'll see later in the book, the cool and important criteria for ranking medical schools and doctors and nurses work perfectly because all the other inhabited planets use augmented intelligence "robots" to do the job of memorizing and analytics, so that their sentient doctors can do their job of healing. In other words, on every other planet, their robots can do what robots do best, which is suspiciously like what Americans used to select and educate doctors to do.

On Earth, in the twenty-first century, these "robots" are starting to appear, although they aren't yet being used effectively for healthcare delivery. Somewhat rudimentary by non-Milky Way standards, they do have cute names like Watson and Parsey McParseface.

Despite this early move to augmented intelligence, doctors in the USA on the planet Earth are still chosen based on those same archaic standards and promoted by the only rankings (USNWR) not endorsed by the Inter-galactic Ranking Council. The president of the Inter-galatic Ranking Council once called the USNWR rankings, in his own language, FUBARM, which roughly translates to "for undergraduates based absolutely on rote memorization."

All this means that rich people in the United States who can afford it will send their children to very expensive courses that teach those rich children how to memorize (like robots) so that they can pass the med-cats right into medical school (without passing Go[7]).

And a student who doesn't own a Mercedes or maybe even a car can be shut out because she didn't get the chance to learn the cool "tricks" to pass the med cats.

In other words, the United States stands alone in planetary healthcare by not letting the robots be robots (and memorize stuff) and the humans be humans (and show compassion, communicate care, and take the time to interpret what the humans under their care are trying to say and need to feel better.)

---

[7] Passing Go is a reference to Monopoly, a board game about real estate that humans universally agree is not fun, but have continued to play for decades.

If you're reading this and you are located in the United States of America, you may be wondering how other planets choose their doctors. It's simple: They show applicants reruns of Earth's Marcus Welby and House and see which one elicits a knowing smile.

2.2 You're a Lot Safer Flying With Me…Than Being Operated On By Me

There is one other thing that seems bizarre to most of the rest of the universe as it relates to how doctors are trained on this planet.

There are these people called *surgeons*.[8] You hope you don't spend too much time with them, although they can fix all kinds of stuff. Problem: They have a unique system for teaching other surgeons new techniques. It's called "see one, do one, teach one."

While it's a catchy phrase, it means that doctor trainees are often practicing things that require extreme skill and dexterity on other humans before they have proven they are competent on inanimate objects.

Inanimate objects are not alive and cannot be hurt. No, they are not cool zombies like *The Walking Dead* but they are made of plastic or silicone, look and act like real human beings and live in houses called simulation

---

[8] Surgeons on earth are doctors who are very good with their hands and who cure things by cutting open people and taking the bad stuff out. They don't smile a lot.

centers. More importantly, they can be fixed easily when a see one-do one trainee cuts the wrong thing (or too much of the right thing). They also do not have inanimate malpractice attorneys.

In every other planet we traveled to, the phrase is *"see one, prove you can do it on something that can be fixed or replaced, and then do one, then teach one after you have proven you can a) do the procedure well and b) communicate with others."*

The above is not a very catchy phrase like "see one-do-one-teach-one" but it is how it is done on every other planet and even in every high reliability organization on the planet Earth. Nuclear energy operators, commercial pilots, and military remote drone[9] operators work extensively on simulators before being allowed to affect indigenous life forms.

Those nuclear power plant operators, commercial pilots, and drone operators (the military ones, not the teenage boys) have to prove their competency every year or two. But surgeons who get to cut into their human counterparts never get their technical competence assessed.

Now that would be fine if surgeons never made mistakes like the robot doctors of the planet Climara who are immediately disincorporated if the operation is not a total success. This rather harsh accreditation standard accomplishes two things, 1) It pretty much makes sure that the robot doctor is competent and 2) there are no malpractice lawyers[10] on Climara.

Believe it or not, what the aliens saw looking at us was that healthcare in the United States in the early twenty-first century had only rudimentary tools to validate surgical and communicative skills.

---

[9] Drones were military aircraft designed to kill other humans outside the United States, without danger to the remote pilot. Through an unusual series of events, they then became the most sought-after Christmas present for suburban teenagers.

[10] Malpractice lawyers are the opposite of surgeons. they do not use their hands and they smile a *lot*. They smile because they get to ride around in fancy cars and their own private jets fueled by a magical potion called "contingency." See next chapter.

2.3 The "C" Word: Forbidden Around the Galaxy (except for the USA)

Last time our friendly aliens visited the Planet Earth, there was this commercial on the T.V.[11]

"If your baby isn't perfect, contact Schwartz, O'Reilly, & Patterson and you may be up for a large cash reward. And it won't cost you anything!" Now that begs two questions, why doesn't it cost anything and if your kid isn't perfect how do you know it's the doctor or hospital's fault?

---

[11] TV is the primary source of entertainment on Earth, although the "internet" is eclipsing it. The primary source of entertainment used to be gladiators fighting and killing each other in a place called the Coliseum.

It even begs the question: "What kid is perfect?"

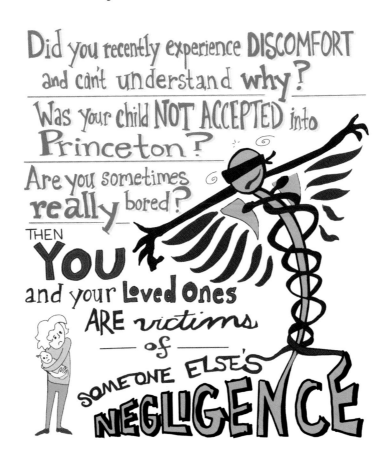

So, let's answer the first question: Why doesn't it cost anything? Contingency is a magical formula that allows people to sue for no money out of their pocket or even bitcoin while plaintiffs' attorneys stand to make lots of money (mostly from the people whose kids aren't perfect, but who only get 50-70% of the pot[12]).

Contingency is not an Earth-wide phenomenon. It is confined to the USA. In other countries on the planet Earth, the people "harmed" get almost all of the money and lawyers don't get to buy their own jets, forcing them to fly on the same planes as doctors.

The "malpractice crisis" has forced many doctors to abandon things like delivering babies or doing difficult surgeries. Doctors that continue to do those things can often be spending a good portion of their salaries on malpractice insurance.

---

[12] "Pot" is the money that gets paid to the winner in "winner takes all" gambling games like poker. In those games the casino takes some off the top. Even they, however, are embarrassed to take 30-40% like the malpractice lawyers.

It also causes a vicious cycle because some doctors practice defensive medicine.

Remember "see one, do one, teach one," the catchy phrase from last chapter that seemed <u>good</u> for medical education but <u>bad</u> for the "ones" on the other side of the table to the surgeon? A similarly catchy phrase exemplifies defensive medicine, "you never get sued for doing a C-section[13]."

That catchy phrase has led to a 45% C-section rate in some hospitals because, while it might not be great for the mom, it takes a lot less time, the doctors make more money, and they can go home knowing they will probably not get sued.

Other planets figured this out long ago. They only spend two hours in a typical hearing arguing about whether these mistakes were about "incompetent doctors" or "greedy lawyers." They recognized that indigenous life forms make mistakes (except for those robot doctors of Climara) and that "bad results" are often expensive and that in some cases, those who have to bear the brunt of the bad results should be compensated.

They also recognized that lawyers are smart, hard-working people who deserve to get paid, but not necessarily paid ridiculous chunks from the money that the other hard-working people who had the bad result (or less-than-perfect kid) received to compensate for that bad result.

On other planets, there is not even a word for contingency. The typical system in the intergalactic council works very well with panels of doctors and lawyers working together to help the patient receive dollars to pay for and in some cases receive extra compensation for bad results.

In fact, most planets got past the "it's about greedy lawyers," versus "it's about incompetent doctors" early in their biologic evolution even before their forebrain was developed. So, the blame game has become an anachronism almost uniformly throughout the universe.

---

[13] A form of delivering humans named after a very strange Roman emperor.

Physicians, lawyers, and patient advocates actually got together to look at the issues and interests involved in medical errors. They realized the issues revolved around reducing poor outcomes through technology and training while allowing patients who have been harmed to have access to a "no-fault" poor result and medical error compensation system.

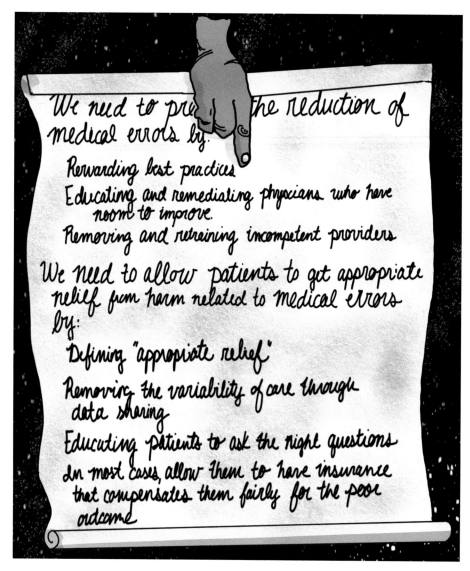

In fact at the two hundredth anniversary celebration of the Euphorian solar system truce between doctors and lawyers, they unearthed (or un-Euphored) the original scrolls that led to that truce:

We need to promote the reduction of medical errors by:
   Rewarding best practices
   Educating and remediating physicians who have room to improve
   Removing and retraining incompetent providers

We need to allow patients appropriate relief from harm related to medical errors by:
   Defining appropriate relief
   Removing variability in care through data sharing
   Educating patients to ask the right questions
We need to allow, in most cases, patients to have insurance that compensates them fairly for the poor outcome.

Amazingly, according to the Euphorian Universal Unity history books, that treaty was "released and sent to all solar systems including the Milky Way." But according to the *Hitchhikers Guide to the Healthcare Galaxy*, the treaty was intercepted and destroyed before it reached Earth by some powerful people on Jupiter who own high end consumer rocket aircraft, fearing that plaintiff lawyers from Earth would no longer be able to afford their rocket ships if the no-fault Euphorian plan was adopted.

So, simply put, defensive medicine doesn't happen anywhere else in the universe. Everywhere else babies are born naturally, not by C-section, (except for Alpha Centauri, where the babies are delivered naturally through an opening in the lower abdomen that eerily matches a human C-section scar).

Doctors everywhere else sleep better at night knowing that a bad result will not result in them being asked to recount five years in the future the exact events of the delivery that occurred at 2 AM five years previously and then try to convince 12 of his/her "peers" (who are not baby deliverers or even doctors) why an electronic fetal monitor deceleration was not harmful to the baby.

In most other planets, there is no need for plaintiff experts[14] who get a chance to say what they would have done if they were present at 3 o'clock in the morning (which is usually not what the doctor did and usually involves a C-section).

On these planets, patients are happy, doctors can practice medicine instinctively, and lawyers on those planets still smile more than surgeons and don't need to use their hands (except to put the luggage in the upper compartment of the commercial flights that they fly on.)

---

[14] A plaintiff expert is a professional whose job it is to make the lawyer happy. He or she is not always an expert at all but is often very good at relating to the the other 12 "non-experts" on the jury who will decide the case.

2.4 It's Hard to Get Someone To Do Something When Their Salary Depends Upon Them Not Doing It

There was a very smart writer who wrote a book called *The Jungle* about the meatpacking industry. His name was Upton Sinclair.

Sinclair once wrote, "It is difficult to get a man to understand something, when his salary depends on his not understanding it."

You're probably asking, "what the heck does this have to do with the mess that is healthcare in America?" Well, everything!

Let's try a MATH problem...
This guy did 100 gall-bladder operations! 20 patients returned with complications.
Doctor Dirtyhands

This guy did 100 gall-bladder operations! 0 patients returned with complications.
Doctor Scrubwell
WHO got PAID MORE?

Let's take Dr. Scrubwell and Dr. Dirtyhands. Dr. Scrubwell and Dr. Dirtyhands have the same amount of surgeries in a year. Dr. Scrubwell's patients have almost no wound infections and hardly any of his patients come back to the hospital to be retreated within thirty days. Dr. Dirtyhands has very different results. Fifteen percent of his patients end up with infections and other complications and have to return to the hospital. Which of these doctors was better compensated? If you said Dr. Scrubwell, you are very smart and very wrong ... and Upton Sinclair was right (both about the meatpacking industry and about perverse incentives.)

Dr. Scrubwell looks sad, because he was an A+ surgeon but Dr. Dirtyhands got to bill insurers, Medicare and Medicaid every time he brought one of the complications back into the hospital. Dr Dirtyhands drives a nicer car than Dr Scrubwell because, even though he wasn't doing such a hot job, he got rewarded in FFS[15] with OPM[16].

You will remember OPM is the first of the things we mentioned at the end of Chapter 1 that don't exist on any other planet.

This man is a hospital CFO in 2007. He is very smart with figures and spreadsheets, and knows initials like EBITDA, CCR, CIPI and CON, most of which have become more important because of OPM. He is reading a spreadsheet. That spreadsheet is the hospital's daily census. What he noticed is that Dr. Dirtyhands is doing more surgery than Dr. Scrubwell. He recognizes that his spreadsheet gets bigger and greener when doctors do more surgery. What he didn't recognize is that Dr. Dirtyhands often operated on the same people that were there yesterday.

Why didn't the hospital CFO notice that? Because it didn't matter financially to the hospital. OPM would pay anytime something was done, regardless of the cause. So if there were a lot of Dr. Dirtyhands with fifteen percent wound infection rates, the hospital's bottom line was not negatively affected. The more the better.

So, until very recently, we paid doctors and hospitals for overutilization.

There was a brief period of time in the 1990s when managed care[17] came into town with gatekeepers and capitation that tried paying for healthcare in a very different way. Managed care tried to make "primary care" doctors the arbiters of payments to specialty doctors. Our alien friends wondered why that didn't work out so hot.

---

[15] FFS stands for fee for service, a bizarre payment model that made doctors using a knife wealthier the more surgery they did, even if some of it was not always necessary. The good news is that this model is an anomaly that only exists in the place called the United States of America on the planet called Earth (similar to the story about plaintiff lawyers and contingency payments in previous chapter).

[16] OPM stands for "other people's money" as in "I made a billion dollars selling hedge funds with OPM."

[17] Managed care: Think Darth Vader in a three piece suit. Managed care was a euphemism for "we will make sure we get you the least expensive alternative and limit your choices."

This is a primary care doctor. She works really hard, is the heart of healthcare, and is often told that she is the quarterback of the system.

This is a real quarterback.

He works very hard, is the heart of the football team, and makes a LOT of money (because he works very hard and he is the heart of the football team).

So, the primary care doctor in the USA makes lots of money since he or she is the quarterback, right? Wrong! He or she makes less than just about anyone (except for maybe pediatricians and psychiatrists) because based on payment alone, people on Earth don't care about babies or brains (unless you are doing surgery on the babies or brains to remove something and then there is plenty of OPM for that).

Well, Managed Care said, "we will change all that, primary care provider, and we will pay you more if you do everything yourself and don't send patients to those expensive specialists that are hungry for more FFS and OPM."

Imagine if a quarterback was paid more if he never handed the ball off to a running back or passed it to a wide receiver but kept it all the time. That would make for a really bad football team (think Cleveland Browns) and it made for a really bad healthcare system. Bottom line for managed care: We paid for underutilization.

You're thinking, "we've tried promoting overutilization, we've tried promoting underutilization and they were both bizarre." Hmmmm. Why don't we promote optimal utilization?

Boy, you are smart. But hold on there cowboy, not so fast. First we have to *define* optimal. Until very recently that has been impossible, like penetrating the Death Star in *Star Wars*. Just as in *Star Wars*, here come Analytics and Decision Support to save the day.

Think of Analytics as Han Solo and Decision Support as Chewbacca (probably without the hair, as it would mess up the algorithms).

Yes, on other planets, Analytics and Decision Support, with a lot of AI[18] sprinkled in, allow teams to be paid based on optimal utilization. On those planets, the "dark side" of Managed Care and FFS were firmly defeated (which obviated the need for any bad sequels).

As of the writing of this book, the place called the United States on the planet Earth (where they still eat a lot of Milky Ways) is furiously working to replicate this model. They have funny names like ACOs and CINs. ACOs are "Accountable Care Organizations," or when they started and there were few rules, they stood for "Anyone Can Open" one. CINs are "Clinically Integrated Networks," but pronounced as "sins." I will leave that to your own interpretation.

OK, so we're making Upton Sinclair style progress, right? First of all, in 2017 thanks to the Vegan movement, there is a lot less meat to pack so *The Jungle* is much less jungly.

And we are starting to come up with non-bizarre ways of paying physicians that would make Upton proud, and all of us healthier.

But alas it is not that easy. Because you forgot our poor quarterback, the primary care doctor. She is still making less money than everyone, except the kicker. So we are getting to the last and most bizarre part of all the messy payment problems in healthcare.

---

[18] Augmented intelligence (think Yoda).

This is a dermatologist in Florida.[19] She knows a lot about the skin. She also knows a lot about money because she gets a lot of OPM transferred to her account every time she cuts out a little mole. If she cuts out two moles, she makes twice as much money. If she cuts out three moles, all of you in unison … .

This line around the block is medical students wanting a dermatology residency.

---

[19] Florida is a state in USA with no income tax, lots of sunshine, and more than its share of hurricanes, sinkholes and other bizarre happenings. People in Florida have lots of moles (the skin kind, not the small mammal variety).

This much shorter line is medical students wanting to be a family doc.

So how do people of USA on Earth fix this? It will require three things:

1. There needs to be less OPM and more HDHPs[20] so that people will demand answers to questions like, "Why are you charging me so much to freeze a little mole?"
2. We need more dermatology training programs. Why don't we have more dermatologists? Because someone in a big office in Washington who runs dermatology training took Economics 101 in college and figured out the supply-demand curve. That is, if the supply of dermatologists is low, the demand will be high and OPM will pour in. Easy change: More dermatologists.
3. The people who pay for care have to wake up and start paying quarterbacks what they are worth, whether those quarterbacks keep the ball or not!

Oh … and one more thing. If you really try to beat Managed Care (Darth) and FFS (the emperor) with Analytics, Decision Support and AI (Han, the Wookie and Yoda), lots of people will tell you that they have already tried. But they haven't.

Finding the flaw in the Death Star wasn't easy, but it was a matter of survival. Getting rid of FFS, managed care, and OPM will be equally difficult (especially without light sabres) but the "force" will get it done if you listen to Yoda saying, "Train yourself to let go of everything you fear to lose and unlearn … all that you have learned!"

---

[20] HDHPs are high deductible health plans or for folks who have them, "Hellish Deviously High Prices."

In the song, *People Get Ready*[21], the lyrics go like this "people get ready … there's a train a'coming … don't need no ticket … you just get on board." The train is not a real train but inspired by the Underground Railroad that led enslaved African Americans to freedom.

Our take away: The passion and heroics that result in freedom are led by "everyday people" (actually another great song, from Sly and the Family Stone).

So, what does this have to do with healthcare? It will need "everyday people" like YOU to break through the "fragmented, expensive, inequitable, non patient-centered and occasionally unsafe" healthcare delivery system. You are the conductor on the train "that's a comin'."

In fact, our inquisitive alien friends recognized this. They said, "we have looked at medical students, surgeons and lawyers, and now when we look at people on the other side, at regular people in distress, people avoiding distress, even people in good health, we know that we/they deserve health care that is thoughtful and really patient-centric (as opposed to the feel-good ads and billboards showing frolicking happy patients that healthcare providers and insurers hire marketers to portray).

This is the "people" side of the story. This is YOUR story and responsibility. So, people get ready!

---

[21] People Get Ready. A great song by Curtis Mayfield and the Impressions released in 1965, the same year Bill Kissick joined a team to write Medicare into law (more about Bill Kissick soon).

3.1 Healthcare's Customer Service Is a Real Turkey

Remember in the first chapter when we talked about Thanksgiving?

This is me on Thanksgiving 1975. Like many teens, I ate too much store-bought turkey, stuffing, pumpkin pie and anything else that my mother put on my plate.

The day after Thanksgiving we got up early, with no breakfast, partly because I had a stomach ache, partly because my dad wanted to beat the crowds at the mall, and partly because I was worried I had forgotten to do my one job, which was get milk at the convenience store.

This is us waiting in a ridiculously long line at the toy store only to find out that the Cabbage Patch Dolls[22] that my sister wanted were sold out.

This is me, that night, still with a bad stomach ache. It was really bad. My mom called the doctor.

---

[22] Cabbage patch dolls: Ridiculously priced funny looking "things." Your ability to snare one before the holidays was a prime measure of whether you loved your kid or not.

This is my mom waiting on the phone for the answering service to pick up.

This is my mom … still waiting for the answering service of our doctor to pick up.

This is my mom, looking angry because the only options she was given for my worsening stomach ache were to go to the ER, tough it out, or see the doctor on Tuesday (apparently my doctor needed five days to give thanks).

Fast forward from 2007. This is now my teenage son on Thanksgiving 2017.

He, like many teens, demanded we go to Whole Foods and then ate too much grass-fed, free-range, non-stressed turkey, pumpkin pie whose roots were fertilized by manure from grass-fed animals, and anything else that my wife put on his plate except for stuffing, because he is gluten-free.[23]

The day after Thanksgiving my son got up early, with no breakfast, partly because he had a stomachache, partly because our Amazon Echo reminded him that he had fallen asleep before finishing the final season of "Game of Thrones" on HBO Go, and partly because he had forgotten to do his one job, which was stay awake while the Instacart guy was delivering the milk.

---

[23] Celiac syndrome was and is a horrible disease that required those afflicted to abandon gluten their entire lives. Millions of young folks in America decided it is "cool and healthful" to abandon gluten, mostly because the people who make gluten-free products have convinced them that it is indeed "cool and healthful."

But despite the stomachaches, we all still went shopping, online that is. This is us, not looking happy, realizing that because we had waited so long, that "the Hatchimals"[24] that we wanted to get his sister were sold out on Amazon and we were getting "schtupped"[25] by eBay with Super Bowl ticket type prices (plus $29.95 in shipping to get it before Christmas).

Back to my son's stomachache. This is my wife waiting on the phone for the answering service to pick up.

---

[24] Hatchimals: Ridiculously priced funny looking "things." Your ability to snare one before Christmas was often the prime determinant of whether you loved your kid or not. Hatchimals were also accused of uttering foul language when they said "hug me" and many people thought that the "hug" sounded like another word that involved a very different physical act.

[25] Schtupped: A Yiddish word originally meaning what some people thought their Hatchimal was saying and now standing for paying five times retail for a stupid doll that my kid will be tired of in three weeks.

This is my wife … still waiting for the answering service of our doctor to pick up.

This is my wife … looking mad because the only options she was given for his worsening stomachache was to go to the ER, tough it out, or see the doctor on Monday (apparently, given decreased reimbursements, alternative payment models and Obamacare[26], doctors now only get four days to give thanks).

Thank you for calling Pleasantville Family Medicine. Please listen carefully to the following 700 options as our menu may have changed…

---

[26] Obamacare: A magical code word now strangely applied to anything that doesn't work in healthcare. It immediately absolves everyone who could actually fix the system from having to actually fix it, because "it's Obamacare's fault."

What's wrong with this picture? Forty years! Forty years! Everything's changed except healthcare!

Look at music.

Look at Television.

Look at Movies. Whoops!

Well, maybe other than the Rocky, Star Wars and Planet of the Apes franchises, EVERYTHING has changed - except for healthcare.

Whose fault is this?

The failure to change falls on the feet of one group. No, not doctors, not lawyers, not pharma, not insurers, not even (dare I say it) Obamacare. It is all *your* fault …. and my fault … and every patient who tolerates service that would yield a one-star on Yelp but that you, I, and everyone else tolerates in healthcare.

So, here's the plan. I want all you millennials[27] and older people who get it (including all of you baby boomers who need lots of healthcare and have learned how to use Facebook to see pictures of your grandchildren playing with your grand-dogs) to rise up and say:

"I'm mad as hell and I'm not going to take it any more.[28]

---

[27] Millennials are usually viewed as young folks who watched boomers, X-ers, and Y-ers mess lots of things up and now don't trust anyone they can't follow with their thumbs.

[28] An iconic line from a 1970s movie, *Network*, quoting a network anchor upset that journalism had fallen to corporate ownership and ratings (which makes one other thing other than Rocky and healthcare that hasn't changed since 1970).

Demand that you get from healthcare everything you can do in the rest of your life. Demand that it not be harder to make an appointment with your gastroenterologist than it is to schedule a trip to the Ice Hotel in Sweden.

Demand that you can get your rehab after your knee replacement at home using virtual technology (the same virtual technology that allows you to defeat Darth Vader on your Playstation 4) instead of going out in the snow to an office and risking needing new knee replacement surgery.

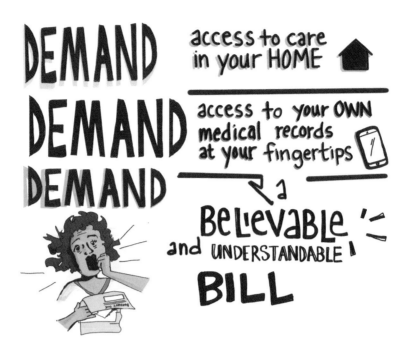

Demand an interoperable medical record at your fingertips so that tracking your cholesterol and triglycerides over the past five years is as easy as tracking how many Star Trek episodes and toys and autographed pics you have ordered on Amazon over the past five years (not that there's anything wrong with that!).

So BASICALLY,...
DEMAND
the same CONVENIENCE
you've come to expect
EVERYWHERE else!

The technology is there. It's your fault! If you demand it, doctors, hospitals and others will realize the gig is up and give it to you. It is economics 101. Oh, and one more thing. Absolutely, ABSOLUTELY, demand a ... BUB!*

* See next chapter

3.2. Hey BUB[29], I Have No Idea What You Charged Me For

This is a bill when you order something from Amazon, Walmart or even Sears and Penneys.[30]

It is believable and understandable. If it was not you would a) raise holy hell and b) never go to that store anymore. If you were really mad, you would probably blast the store on Twitter to your 90 followers (unless you were Kim Kardashian, in which case 4,598,245 people would know that store stinks).

This is a bill when you go to your local hospital's emergency room. It is as unintelligible as if the bill came from Jupiter.[31]

---

[29] **BUB: Believable Understandable Bill … or the Holy Grail, as we call it in healthcare.**

[30] Sears and Penneys: Two stores that may not be around by the time you read this book because they forgot what consumers need.

[31] Jupiter is a planet in the same Milky Way galaxy as Earth. It is unclear if Jupiterians get believable understandable bills in their emergency rooms. If they do, please ignore the slight. Although it is true that Earth people have a long history of questioning the intelligence of Jupiterians, eg. "Girls go to Venus to get more genius, boys go to Jupiter to get more stupider!"

Why can't hospitals and insurers send you a believable, understandable bill like Amazon does when you bought the entire collected episodes of Gilligan's Island?

That may be one of the great questions in all of Earth - both why you would want to watch all of the episodes of Gilligan's Island, but also why that bill will be so much more understandable than that of your healthcare provider.

The other great questions in all of Earth include, a) why did they do a remake of Planet of the Apes, b) why does Taylor Swift still have boyfriends when they know she is going to trash them spectacularly in her next song, c) when will Harrison Ford star in "Indiana Jones and the Legend of Bingo Night"?

My best guess is healthcare is BUB-free because of one of the themes of this book. Repeat after me. OPM. (Other People's Money.) Other people pay the bill. Think of it this way: Imagine if you checked out of Whole Foods, got an unbelievable total, and when you asked to see the receipt it was as undecipherable as a Jupiterian book. What would you do?

This is what you would do: You would beat your fists, you would scream and holler. You might even have a tantrum and throw your cage free, unstressed chicken eggs out of your fair-trade, ecologically-sourced, reusable bag.

But when you get an astronomically laughable, ludicrous wacky hospital bill, what do you do? You shrug your shoulders and smile. Why don't you act like a wronged Whole Foods customer???? Because at the supermarket it is your own money. In healthcare, traditionally, it is OPM.

So, once again it's *your* fault. But just as children's books and movies always leave you with a moral of the story, e.g. kindness is king, don't give up when things look bad, don't trust your stepsisters and most importantly it might be "Frozen" but people are "Happy" when a film grosses over $3 billion, we too have a big take-away. *Bless This Mess* offers this moral:

Every time you get a hospital bill, pretend it is your own money!

Then walk into the hospital administrator's office (you will know it's the hospital administrator's office because it is always on the highest floor and is surrounded by lots of people who will give you lots of reasons why you can't go up there). Hint: For fun, bring a friend with a camera.

Now the fun part begins. Ask the administrator to explain each of the undecipherable, duplicative, crazy, nonsensical items on the bill (preferably with the camera rolling).

Here's what you will find: That most of the charges were never meant to get paid because they were part of a contract with the people that pay the bill, the OPM people.

Is there hope in this depressing story? The hope is a combination of two things: The rise of millennials and the rise of high deductible health plans (HDHPs).

As millennials need health care, they will increasingly demand bills that they can view on their iphone XVs or their Samsung Galaxy 18ss (assuming they have not caught fire), with explanations that are written in 140 characters. Further, as baby boomers need healthcare in order to live forever, they will start to act like millennials (which also means paying more of their bills and using less OPM).

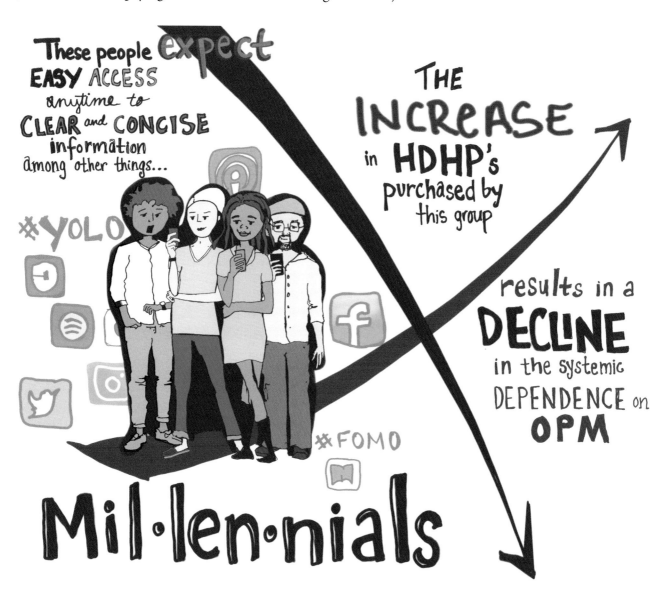

And the HDHPs will defeat the dark force of all healthcare being paid by OPM. That HDHP-OPM heavyweight fight will (after some pain and suffering) result in a new day with new rules, and…

… hospital bills that look more like this:

3.3: Death, Dialysis and Decisions That Have To Be Made (even in Alaska)

Another thing Yoda said: "Death is a natural part of life. Rejoice for those around you who transform into the Force. Mourn them do not. Miss them do not."

Clearly Yoda never spent any time in the ICU of a hospital where often the only discussion is about how long that patient can stay "alive" (as defined by the fact that breathing is occurring with assisted technology and that there is some minimal activity in the brain).

Why do other planets have a more accurate definition of when it is time to "move to the Force" if you are a Jedi, or "Heaven" if you are a Christian, or "Jannah" if you are a Muslim, or "born again into the flesh of another body" if you are Isaac Hayes or Tom Cruise?

The answer is complex. Partly because in the bizarro-land of OPM and FFS, everyone is incentivized in the USA on the planet Earth to keep humans alive well past their "expiration date."

So, here's a little test to see if you've been reading. Your great uncle Charlie is 94. He's disoriented and non-communicative, just had his third stroke and is alive only by the definition above, but goes into renal failure in the USA. What happens? (Answers below.)

a. The nephrologist will most likely recommend dialysis because "you probably want your great Uncle Charlie to 'live longer,' don't you?"
b. The hospital will encourage dialysis because "you probably want your great Uncle Charlie to 'live longer,' don't you?"
c. You will go along with the dialysis because it sounds right that "you would probably want your great Uncle Charlie to 'live longer,' don't you?"
d. Thanks to FFS, the nephrologist and the hospital will get paid well for the dialysis and you will pay virtually nothing, thanks to OPM.
e. All of the above.

Answer: e, because a, b, c, and d happen to be true … and "all of the above" is almost always the right multiple choice answer when the person devising the test is trying to make a point.

If you got the right answer, congratulations. If not please re-read this book (or at least look at the pictures) and try again. OK, one more test.

Your great uncle Eilrahc is 94. He's disoriented and non-communicative, just had his third stroke and is alive by the definition above, but goes into renal failure *on any other planet* in the universe (or any other country on the planet Earth). What happens? (Answers below.)

a. The nephrologist will most likely recommend dialysis because "you probably want your great Uncle Charlie to 'live longer,' don't you?"
b. The hospital will encourage dialysis because "you probably want your great Uncle Charlie to 'live longer,' don't you?"
c. You will go along with the dialysis because it sounds right that "you would probably want your great Uncle Charlie to 'live longer,' don't you?"
d. Thanks to FFS, the nephrologist and the hospital will get paid well for the dialysis and you will pay virtually nothing, thanks to OPM.
e. None of the above.

Answer: e, none of the above, partly because every other country on Earth as well as every other planet in the universe recognizes that when a healthcare system is going to provide access to everyone and must deal with limited resources, then equitable, just and often tough end-of-life decisions must be made … and also because when you are taking a multiple choice test and the correct answer to the previous question was all of the above, "none of the above" is the right answer to the next question 76.2% of the time.

This is Sarah Palin.

She is not Yoda but she was almost vice president of the United States. She is a funny lady, and says things like "the only thing that stops a bad guy with a nuke is a good guy with a nuke." When she was a vice presidential candidate, Sarah Palin said, "the America I know and love is not one in which my parents or my baby with Downs Syndrome will have to stand in front of Obama's 'death panel' so his bureaucrats can decide whether they are worthy of health care."

Now while it's true that sometimes it is difficult to separate the funny Sarah Palin from the "heartbeat away" Sarah Palin, that death panel statement was both the PolitiFact[32] "pants on fire" quote of the year and one of the reasons that the Affordable Care Act did not deal with end-of-life issues. This had consequences: It immediately meant that that ACA would indeed not live up to its name of being "affordable."

So, how does end of life work on those other planets?

There is a lot more thought given to what "life" is both by the people involved (if they are able) and by their loved ones, and more thought about when it is time to go off to the Force, to Heaven, to Jannah or "into the flesh of another body."

And there is an understanding that if everyone is to get healthcare, difficult decisions and compromises must be made, especially as to how much money is spent fruitlessly in the last several weeks of life.

But what they don't do is play with FFS or OPM because they recognize that if we frivolously spend so much money on the last weeks of life, then the money will not be there for keeping the rest of the population healthy.

So, as an Earthling named Ben Franklin once said, "In this world nothing can be said to be certain, except death[33] and taxes[34].

---

[32] PolitiFact: a truth-meter from a newspaper in the previously mentioned state of Florida where there are lots of hurricanes and dermatologists and attempts to differentiate truth, fakish news and really, really fake news.

[33] Unless you get cryogenically frozen and then you are just on hold.

[34] Unless you are our current president of the United States and then, according to MSNBC, we can only guess.

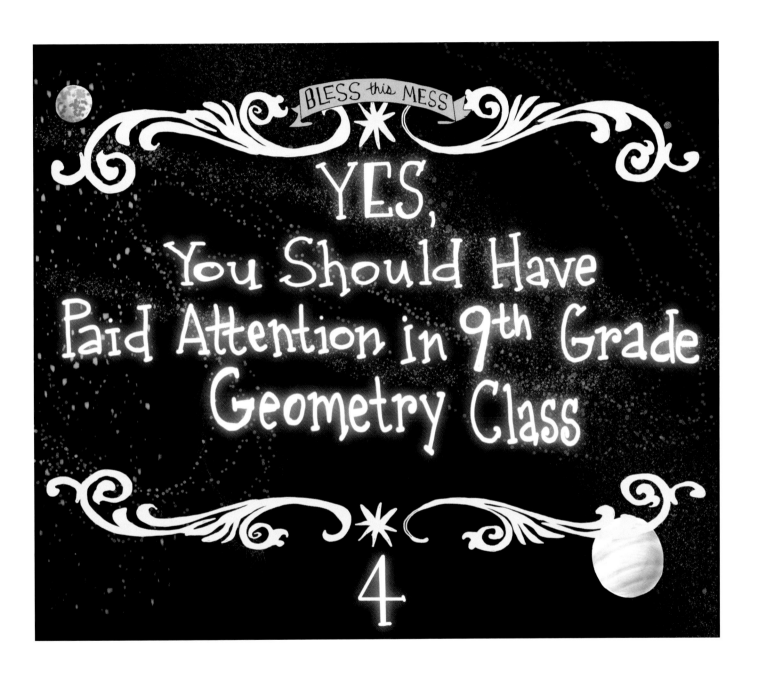

YES,
You Should Have
Paid Attention in 9th Grade
Geometry Class

4

This is Euclid. He was born in 300 BCE. He only has one name like Madonna, Bono, Pink, and will.i.am.

Euclid did lots of cool things including naming the elements - of mathematics. So if you live in Buffalo and are complaining about the "elements," don't blame Euclid.

He also created the framework for geometry on Earth and we now call it Euclidean geometry (so that every 9th grade girl or boy has someone to blame for how boring that lecture is).

Remember 9th grade. No, not your first kiss or the fact that you didn't make the basketball team and had to become equipment manager.[35]

I mean 9th grade math. Where you had to buy a protractor. [36]

Here is a geometry test.

This is a (pick one):

a. Square
b. Rectangle
c. Triangle
d. Rhombus (even though I know it isn't I always wanted to give that answer)
e. What does this have to do with healthcare?

---

[35] See Klasko—autobiography

[36] A protractor is a measuring instrument, typically made of transparent plastic or glass, for measuring angles. It is also the most worthless thing you ever bought (unless you have the VHS of "Rocky 6" or the CD of the Milli Vanilli reunion).

If you answered c, d, or e you deserve a lot of credit.

If you answered a or b, please stop reading this book and immediately pick up some lighter reading.

If you answered c, congratulations you passed 9ᵗʰ grade (you still won't make the basketball team).

If you answered d, I'm glad you got to say rhombus. Now hurry up, you're late for Comic-Con.

But if you answered e, you are allowed to read on.

Alright (e) answerers.

This is what it has to do with healthcare. There are many villains involved in our story about the "mess" of healthcare in the United States. And there are many reasons of how and why we got here (more on that later). But the Voldemort of shapes … excuse me … the polygram which shall not be named … of healthcare policy, is the iron triangle. The iron triangle was written up in a great book in the 1990s.

Bill Kissick (he is not Voldemort, he was actually a great and brilliant professor at Wharton[37]) wrote a book called *Medicine's Dilemmas: Infinite Needs Versus Finite Resources (Yale University Press 1994).*

Sound familiar?

For those of you in ninth grade, this is a good reason to not sleep through geometry class. OK, I'll refresh your memory. A triangle is made up of 180 degrees - when you add its angles together. If you increase one angle, you have to decrease the other.

Stop nodding off. This is important!

In his 1994 book, Kissick wrote that there is an iron triangle in healthcare. And thanks to Euclid, we know that if we increase one angle we have to decrease another. (Don't you wish you had paid attention in 9ᵗʰ grade?) And those angles have to equal 180 degrees. The sides of the triangle in healthcare are access, quality and cost. So if you increase access, you have to decrease quality or increase cost. If you increase quality, you have to decrease access or increase cost … get the picture?

So what does this have to do with our problems today? Well, as Louis Brandeis[38] said, it's hard to argue with "the relentless rules of humble arithmetic."

Translated: If you mess with math or Euclid, you're in deep do-do.

In President Obama's case, and the Democratic Congress, they ignored that pesky iron triangle and played dangerous games with non-Euclidean geometry by saying "we will increase access, increase quality and decrease cost in healthcare."

In President Trump's case, and the Republicans, they ignored that pesky iron triangle and played dangerous games with non-Euclidean geometry by saying, "it's going to be beautiful, terrific, unbelievable and huge," but really meaning we will increase access, increase quality and decrease cost in healthcare.

---

[37] Wharton is the Hogwarts for masters of the Universe, some of whom really are wizards like Bill Kissick, others of whom end up in the dark arts (see Goldman-Sachs).

[38] Louis Brandeis: a Supreme Court justice from the early twentieth century who had a college in Boston named after him that is not named Harvard or Tufts.

And while Bill Kissick did not draft Obamacare or Trumpcare, Bill did know geometry and he did know healthcare when he helped draft Medicare.

He said, it's simple math … and geometry. The only way to get around that iron triangle is to ignore Euclid and make your own geometry.

What it meant back in 1994 is that if anyone tells you they are going to increase access, increase quality and decrease cost without fundamentally disrupting the system and causing pain they (what's that PolitiFact word) have their "pants on fire."

Well, at your next cocktail party and someone asks what's going to happen to healthcare in America just rub your chin and say, "Ah Euclid, ah Kissick." Either someone will think you are brilliant, or tell you that you need to stop drinking.

Because if you really do want to look smart at a cocktail party and not sound too partisan, here is what you could say. The Affordable Care Act did exactly what it was meant to do!

Smirk from the Democrats

Hushed surprise from the Republicans

It gave everyone access to a fundamentally broken, fragmented, inequitable, and occasionally unsafe healthcare delivery system, ignoring the iron triangle of access, quality, and cost and "hoped" that it would be alright because the healthcare system would transform itself.

It didn't!

And part of the reason it didn't is that we ignored Euclid and Kissick. We didn't go through the hard work of getting it right.[39]

---
[39] See Nancy Pelosi and Mitch McConnell quotes circa 2010.

When it came to Obamacare, we didn't want to mess with:

Pharma prices.

Malpractice insurance - see greedy lawyers, incompetent doctors chapter.

Salary gaps - think dermatologists and quarterbacks.

Insurers - with words like medical loss ratio[40] and adverse selection.

End of life.

Healthcare disparities - where your zip code is more important than your genetic code.

Patient responsibility.

So, we have a very messed-up looking triangle.

Which made a lot of people unhappy who didn't like either their geometry or their healthcare messed with, which led us to a president[41] who claimed to understand healthcare, eg. a real estate developer who figured out that "healthcare is complicated."

---

[40] Medical loss ration: sounds like decreasing it should be a good thing (which it is for insurance executives,) but what it really means is how much of the money that the insurers are taking in that they actually have to spend on care for patients. Hmmm.

[41] While President Trump never actually said that Trumpcare would increase access, increase quality and decrease cost, he did say it would be terrific, unbelievable, awesome and huge. My guess is that both Bill Kissick and Euclid would agree that those statements defy any geometry they know.

Who knew?

Here's the honest truth.

The ACA did what we asked it to do.

But the iron triangle was ignored.

And healthcare didn't transform.

Let's review the culprits:

1) OPM

2) PACS[42]

3) PEOPLE AND CORPORATIONS doing what is in their best interests (imagine that) and hiring lobbyists to make sure that legislators really, really believe them.

That includes:

Pharma
Insurers
Specialist Doctors
Lawyers
Hospitals
Nursing homes
Electronic medical records companies

… and on and on and on.

And you … yes you (do not look around the room).

You want it all. You want your family to have it all for a minimum amount of your own money but you don't want to pay a lot of taxes to have other people have it all … and you don't want anyone limiting your options. Sorry to be harsh, but if I'm asking everyone else to look in the mirror, you are going to have to also.

Euclid, Kissick, and the American electorate don't like when people break the rules of geometry so they took it out on the architects of Obamacare and elected Trump.

Happy or not, Mr. Trump is now President Trump and Obamacare will no longer be Obamacare. That's the easy part. Remember when you spent all day building a sand castle on the beach and your stupid older brother easily knocked it down?[43]

---

[42] PAC MEN and WOMEN: No not the game where those cute creatures eat special pills to become stronger (sort of like some professional sports players) but political action committees (PACs) which gobble up lots of OPM to get folks in the Congress and Senate to do what their conscience tells them not to do.

[43] See Klasko autobiography

Remember how hard it was to rebuild? Remember how you hit your brother with the pail and YOU got grounded.

Well, that's the problem. It is one thing to kick down the sand castle called Obamacare, but it is harder to build a better one, especially when the other party would be happy to throw sand pails at you while you're trying to do it.

If this algebra and geometry lesson is too difficult, a) you may need to be on drugs for your short attention span and b) I'll make it simple.

Not enough people in Washington want to deal with the core issue, which is having accessible healthcare for all. And there is no way that can happen without disruption and some pain.

We have $1.00 to pay for $1.25 worth of stuff in healthcare.

The only difference between the Democrats and Republicans is that the Democrats want to give everyone a quarter to cover the difference. The Republicans want to cut out certain people to get the cost down to $1.00.

But no one is yet addressing the disruptive, innovative and sometimes painful changes that need to happen in order to make healthcare a right for all.

Impossible you say?

Tell the Tarvosians, Vendrizians, Roya Vosans, Vovidons, Amisists, Terronda Prime citizens (they are very formal), Criusans, Tandrans, Octavians or Nyotans if it's impossible. They will laugh at the notion (at least the ones that have mouths) because they have all done it.

So read on, interplanetary healthcare travelers!

Now you see where *Bless This Mess* comes in. It's time to figure out how healthcare delivery in the USA, once a cosmic mess, could become an intergalactic model.

President Trump has surrounded himself with some excellent generals, politicians and folks who fly around the country in private planes to help him.

We have gone one step further. We have assembled the greatest hits of all the ideas across the ENTIRE healthcare universe that could be used to start to rebuild the castle in a way that would make Euclid, Bill Kissick, and maybe even Jill Stein[44] smile.

Since there are 7,865,904,087,234,576 planets in the universe (by last count) we have asked Krik Niatpac, the publisher of the *Hitchhikers Guide*, to choose ten planets that will have the most utility for those of us stuck in the mess that is healthcare in the beautiful country of the United States in the beautiful planet Earth.[45]

TARVOS

In the country United States of America on the planet Earth in the galaxy Milky Way,

hospitals are frightening places where you cannot easily sleep, get good food, or even find a little entertainment. Sound minor?

It turns out sleeping, eating and thinking drive recovery and health.

On the planet Tarvos, their healthcare credo in Tarvosian is "caviara, zzzzzzz, and ntflxia," roughly translated into English as "We recognized that most people that end up in a hospital still want to eat well, be stimulated and be able to catch up on House of Cards or Downton Abbey." So, other than the medical part of their care, they provide on-demand food by partnering with restaurants throughout the area that patients can access when they want it, just as they can at home. Tarvosians recognized that on their planet, having people and places with the skill set to drill holes in Tarvosian brains to remove a tumor is not necessarily compatible with running an awesome kitchen. Technology on Tarvos also got to the point where great wholesome food could be delivered

---

[44] Jill Stein. The candidate that got just enough votes (in states that voted our current president in) to ensure that Hillary Clinton did not become president.

[45] For another guide to alternative ways to do important things, read *The Phantom Tollbooth*, by Norton Juster, or my take on that theme, *The Phantom Stethoscope*, by Stephen K. Klasko and Gregory P. Shea, way back in 1999.

in an easy manner. They also make the hospital room just like home with nice televisions and good music. In Tarvos, patients no longer view hospitals as scary places but as community centers of well-being.

From the ambassador of Tarvos: "We recognized that healthcare and health are two different things. Getting better requires more than pills and surgery, and since robots perform most of our surgery there are no egos in the way. But we also recognized that with some creativity, health systems can partner with hotels and other places that do hospitality right and create healthier, happier patients without costing more money."

When asked how they got their hospital staff to recognize that change of culture, he laughed, "we got tired of handling all the complaints about food in our hospitals. So we mandated that every hospital manager has to eat at least 12 meals a week in a hospital room with hospital food." Tarvans eat seven meals a day. "Amazingly patient assessment of hospital food has skyrocketed on Tarvos since that rule was implemented."

VENDRIZI 3

In the country United States of America on the planet Earth in the galaxy Milky Way, we refuse to borrow an idea even from another country on our OWN planet.

In Vendrizi 3, they search all planets to look at better care and never use the term "alternative healthcare" for modalities used to treat chronic diseases that are utilized by providers on other planets. It seems that in the United States, anything that isn't as it has always been done is immediately dismissed. On Vendrizi 3, whenever they find a planet that does things better than their doctors, they call it "tht'sobv'ios" which means "why didn't we think of that?"

On earth, we don't even have to look at other planets. In the country called the United States, ayurevedic medicine and acupuncture are considered alternative (or worse). In many parts of Asia, they are the modalities of choice. So why not just bring the planet Earth together, and create international physicians and nurses? Why not have schools that teach the best of both worlds? And why not recognize when pills and surgery don't work there might be a better way?

From the ambassador of Vendrizi, "It is simple, we look at outcomes across the planets and we welcome other doctors and nurses. We have simulators with cross planetary standards that if you pass, we welcome you and your different ways of making patients healthier."

Readers, I know what you're thinking and you're right…. "tht'sobv'ios"!!

ROYA VOSA

In the country United States of America on the planet Earth in the galaxy Milky Way, we allow professionals called doctors to believe they are "captains of the ship," and act essentially alone in saving the lives of patients.

In the small planet of Roya Vosar, they recognized a long time ago that what really matters in choosing physicians and nurses is that they can work as teams, that they can observe and communicate and that they can work well with the augmented intelligence robots. It amazes them that in the United States we still try to choose our doctors as if they were practicing to be robots.

From the Head of the Vosan Health system, "On Roya Vosa, there are two types of Vosans … those who can memorize anything put in front of them but can never read between the lines, that is, understand what someone is <u>really</u> saying to them; and those who are amazingly intuitive and can feel what someone else is feeling. We learned long ago that both are needed and they work as teams."

"It seems like that potential now exists in your world. It's strange to us that humans think they can always possess both, but apparently that is very rare. Fortunately for you, your technology has developed to the point that just memorizing things can be done by non-sentient beings and therefore you have a golden opportunity to change how you select and educate all healthcare professionals."

"By the way, we teach all our healthcare professionals starting their first year <u>together</u>. There are no educational silos, so that they understand how important they are to each other. In fact, we put them in study groups with no more than two doctors, nurses, etc. It is amazing how quickly they learn to be teammates before the virus of arrogance has set in!"

YOVIDO

In the country United States of America on the planet Earth in the galaxy Milky Way, we want you inside **our** buildings, **our** hospitals and clinics and offices, before we cure or help **you**.

We spend lots of money doing advertisements on "Fox and Friends[46]" or "Morning Joe[47]" to have you choose **our** place to go when you are sick.

We appear afraid of shifting healthcare to your neighborhood, your home or wherever YOU are.

On the planet Yovido, they have moved healthcare from hospitals to outpatient centers to the home, where now most of healthcare occurs through telehealth, machine cognition and other technologies (partly because Yovidans have no arms or legs which makes it hard for them to get around on those cold winters in Yovido).

From the Yovidan healthcare secretary, "It seems as if in the United States you talk a lot about population health, but it is from the point of view of a hospital. In a recent visit to Earth, we saw these strange organizational creatures called ACOs, which someone told us stood for accountable care organizations. We assumed there was a language barrier because most of them seemed to be run by groups of hospitals that did not seem to be accountable. It was unclear to us if the humans running these large inefficient hospitals understood the hospitals could not survive in their current form if care really was disrupted and transferred to the home."

"On Yovido, we pay all healthcare providers based on a yearly assessment of how healthy the population is. The healthiest hometowns result in a bonus for the doctors and nurses and a tax rebate for all the citizens. Not only that but half of the incentive pay for anyone running a healthcare organization is how the population for the entire city or area is doing healthwise. It is amazing how often these healthcare leaders talk to each other, which does not seem to be occurring in your country."

---

[46] Fox and Friends. The reason our current president got elected according to MSNBC.
[47] Morning Joe. The reason some people who didn't vote for our current president want him impeached according to Fox News.

AMISIS

In the country United States of America on the planet Earth in the galaxy Milky Way,

people who are poorer, or in a minority, or less able to travel to our clinics, get healthcare that is often worse, less evidence-based, more judgmental, and this contributes to shorter lifespans.

On the very serious planet of Amisis, the population actually comes from two different planets, but even so they do not tolerate any disparities. One hundred Earth-years ago, the residents of the planet Orgullo destroyed their own environment - massive building schemes, climate change[48], burning fires, weapons of mass destruction freely used. They destroyed their own planet. Chastened, a small group of Orgullians arrived by spaceship on Amisis and settled within their cities.

But the population of Amisis made a primary decision, namely that Orgullians would get exactly the same healthcare as residents of Amisis[49].

According to an Amisisan official, "If there is any city on Amisis that has more than a five year difference in life expectancy between areas of the city, we fire the mayor, public health officials and CEOs of all the health systems in that city. It took less than a hundred years for the systems and the local health bureaus to work together to understand that eighty percent of a community's health does not occur in the doctor's office or the hospital. Now those CEOs are chosen based on their ability to keep the community healthier. They can be original residents of Amisis or Orgullo. It doesn't matter. What matters is that every child has an opportunity to lead a healthy life."

---

[48] Climate change. Thought to be a hoax by many Orgullons, who as the planet was being destroyed, said "Whoops, I guess I was wrong."

[49] For another planet that struggled with integrating populations, see the trilogy *Lilith's Brood*, by Octavia E. Butler, 1987-1989.

TERONDA PRIME

In the country United States of America on the planet Earth in the galaxy Milky Way, we pass along surgical knowledge with a simple formula: "See one, do one, teach one."

On Teronda Prime, "we are amazed that in USA that you did not use technology to ensure that every surgeon can objectively prove appropriate competence and confidence for the procedures they perform. Hospitals and doctors are penalized if they cannot validate that their surgeons are 'competence certified' objectively in each of the procedures they perform. It only took 7 TYs (Terondan years[50]) to have the hospitals that failed to do that close and become storage sites for online shopping.

---

[50] A Terondan year is of variable length determined by a group of elders who decide it will be a new year when the bizarre things that have happened are eclipsed by positive developments. It makes for a happy new year and the New Year's toast is always "let's hope for a short year!"

CRIUS

In the country United States of America on the planet Earth in the galaxy Milky Way,

it appears hospitals are designed deliberately to make them confusing, intimidating and with only a nod to different languages.

On Crius, "We hire random people of average intelligence and good eyesight (Criusians have 5 eyes) to navigate the halls of random hospitals and find various important areas such as nursing stations, ERs, ORs etc. If it takes more than an earth minute to find exactly where to go, that hospital or healthcare facility is closed until they get the appropriate signage so as not to confuse people."

"There is also a language diversity component. We hire surveyors that speak the top three languages spoken in any service area. If any language is not understood, we send notices to everyone in the community that speak that language in their native tongue that they should avoid that hospital and its providers. Pretty effective."

TANDRA

In the country United States of America on the planet Earth in the galaxy Milky Way, hospitals work hard to ensure doctors will send new patients to their institution.

On Tandra, it's the patient's eyes that count. The health minister for Tandra was proud of the fact that "on Tandra it's all about how you treat one another so it should be no surprise that, thanks to new technology that allows us to film hospital interactions directly from the patient's vantage point, literally through the patient's eyes, we now take ten random samplings from each hospital and have other hospital CEOs watch the interactions and grade each from one to ten based on the 'would I want to (or not want to) be treated that way' scale. Half of a hospital CEO's salary is based on that scale."

OCTAVIA

In the country United States of America on the planet Earth in the galaxy Milky Way, every hospital gets a blue ribbon. The hospitals advertise their blue ribbons on things called billboards. In fact the hospitals spend millions of dollars to "brag about" their rankings and blue ribbons. All children also get blue ribbons, but they just stash their blue ribbons in their boxes of beanie babies.

Octavia is a planet of merchants and salespeople. You can't go two feet without some advertisement about a doctor or hospital getting some award (somewhat like Miami).

From the Octavian minister of health and marketing, "we got tired of passing three signs in a row that said three different hospitals were #1 in heart/lung/orthopedics or anything else that provided a margin for the hospitals. So now we hire recent graduates to read all the hospitals billboards, news ads, social marketing and web claims and investigate them for any semblance of truth and/or relevance. Any health systems that grade "below zero" must put a sign on all their marketing over the next year that is the equivalent to your 'pants on fire" designation in the United States.

NYOTA 5

In the country United States of America on the planet Earth in the galaxy Milky Way, even if you can understand the math of your hospital bill, you will never believe it.

As we end our hitchhiking through the galaxy, perhaps my favorite interplanetary lesson comes from Nyota 5. They passed a resolution in the Nyotan planetary council that throughout the land if a patient gets a bill from any healthcare provider that it (Nyotans do not have genders) cannot read, the bill is sent to a 15-year-old of average intelligence. If he/she/it cannot in five minutes say what was done and what it cost, that patient no longer has to pay that bill. Hospital bills have gone down from an average of 27 earth equivalent pages to 1 Nyotan page.[51]

---

[51] Since Nyotans are small, most of their paper is 3" x 5" which makes it all the more impressive.

So, whether you loved Obamacare or thought it was an evil plot to take away your freedom and life as we know it (yes some people actually believed that) it is clear that we can overcome the iron triangle with some disruptive thinking.

President Trump, I believe it would be a great idea to travel to some of these planets to see how well these transformations have worked. While I know that Air Force One does not yet leave the planet, I believe the Roya Vosarians would gladly let you hitchhike on one of their royal rockets. By the way, if you've worked out those conflict of interest issues, the rumor is that they would love a Trump Tower in Roya City!

The Prime Disruption

...or...

Captains Say the Darndest Things

6

In the first chapter, I told you that aliens occasionally looked at this country called the United States. We don't even know when they are looking. But we know they can see our inefficient, inequitable, fragmented system.

SURPRISE! In 2035, those very same aliens invited the United States to join the prestigious Intergalactic Health Council. It was one of the great turnarounds in interstellar history.

Thanks to many of you reading this book, the United States rapidly developed an ideal healthcare system, based on lifestyles of happiness, on equity, on community access to the very best medicine, on healthcare based at home, and on teams of professionals who love their work and enjoy using augmented intelligence to do things we didn't dream of in 2020.

By 2035, patients were seen as critical members of both medical teams and scientific research teams.

How did this happen? What created a "prime disruption" so great as to reverse the mess?

When I was growing up, anything that didn't make sense to me somehow got resolved on the set of a television show called *Star Trek*. That's where we saw everything from inter-racial, or inter-planetary dating to peace with your neighbors to how humans and other sentient beings deal with technology (that may be smarter than them.)

Is it possible that *Star Trek* fixed healthcare in the United States? Much like *Star Trek IV: The Voyage Home* saved the whales?

How would that happen? We would like you to visualize the moment. This little detour may sound like fan fiction, but it may also be how by 2035 the USA was honored to join the Intergalatic Health Council.

Think of the set of the Starship Enterprise. Got that? Now, up to the bridge walks Deanna Troi and Bones McCoy. For those of you either too young to remember or who had a life in the 70s, 80s and 90s and did not watch *Star Trek* or its reruns, Counselor Troi is a Betazoid from *Star Trek Next Generation* who can read minds and Bones McCoy is a doctor on the original *Star Trek*.

In our scenario, our favorite Star Trek heroes appear from the future to tell us today what we need to do.

Using her space-time continuum bending projector, which shows on the smartphones of everyone in the USA, Counselor Troi speaks first:

"We were monitoring the time space continuum and happened to survey Earth in 2018 and boy what a mess. Really, what a mess. But when we got to the USA, we were amazed at the struggles to come up with a model that allowed healthcare to be a right and not a privilege."

"So, I brought a doctor from the future, Dr. McCoy, to help us understand what the United States needs."

"Damn it, Deanna," groused McCoy. "I'm a doctor, not an MBA, and certainly not a professor."

Troi laughed, "That's OK. I brought some pretty good leaders to help."

And sure enough, in walked Captain Kirk, the visionary charismatic leader from the original *Star Trek*[52]; Captain Picard, the ultimate manager, Number One from *Next Generation* (STNG[53]) the quintessential number two person in an organization; and Spock, the logical Vulcan.

Kirk began: "We are here to tell you everything you need to know so you can get started on this healthcare transformation thing. Bones, Spock, you're joining me on an away team to USA in 2017." Picard shot in, "Captain with all due respect, that would be a clear violation of the prime directive.[54]"

Number One chimed in, "Look we have investigated this and we have discovered that this era is the nexus in the time space continuum that will determine whether American healthcare truly transforms, so let's just talk to these folks and give them some of the tools they need."

Troi added, "I am sensing confusion and pain among the leaders of the country at this time. They have even come up with bizarre and funny names to solve their crisis. There is the 'Affordable Care Act that isn't very affordable for many'. Then there is the 'American Healthcare Act which worked to deny healthcare to many,' and finally the 'Better Care Reconciliation Act' which did nothing to define better care and really did not reconcile a lot."

Picard smirked and quickly glanced toward Kirk. "As the acknowledged Starship Enterprise captain with the greatest leadership skills, it gets down to a very simple equation. Every once in a while an industry goes through a seminal event and that is exactly what happened to healthcare in 2017. It was an impossible equation and it needed disruption but there was no urgent event to get everyone off their own little part of the puzzle and come to a mutually beneficial solution."

---

[52] * Bones McCoy, Captain Kirk, and Spock: Iconic figures from the original *Star Trek* franchise. Captain Kirk was a visionary Starfleet leader who often left the managing to others. Bones McCoy was the ship's doctor with a tricorder, an augmented intelligence computer which did most of the heavy diagnostic work. He often mixed it up with Spock, a half Vulcan who believed that the human component of leadership was distracting, not dissimilar to our means of selecting applicants for admission to medical school.

[53] Captain Picard, Number One, and Counselor Troi: Iconic figures from *Star Trek: Next Generation*. Captain Picard's leadership style during a crisis was to call his crew to order, ask their advice, listen to each person and then make a decision based on that input. In contrast, Captain Kirk gave great speeches, made decisions on his own, and told his crew what to do. Number One was Captain Picard's right hand man. (Interesting that he was not called #2.) Counselor Troi was a betazoid, an alien culture with the cool ability to read people's minds and emotions.

[54] The prime directive was the guiding principle of the United Federation of Planets which prohibited Starfleet personnel from interfering with the internal development of other or past civilizations.

Number One took over. "In round terms, the computer tells me, about twenty percent of the USA population cannot get the care they need to live a healthy life."

Spock couldn't help himself, "Actually 21.112% of the population, and that represents a difference in life expectancy compared to other humans of approximately 14.87156… ."

Number One got back to his point, "What if we find some unknown author and have him write a book called *We Can Fix Healthcare in America: The Future Is Now*" and just give him the 12 precepts behind the Starfleet and federation health plan?"

Counselor Troi stepped in, "Excuse me Captain but how do we know leaders will embrace the precepts?"

Picard continued, "Leading amidst the swirl of change requires a range of behaviors carefully matched to the situation and the people. Those kind of creative partnerships and lack of certainty are not what physicians studied in medical school. It's not what the white coat brought them in yesterday's healthcare."

"Most of your current physician leaders such as deans, chairs, and chiefs did not develop in a world of service lines and profit/loss statements let alone with multiple metric dashboards, risk management, and competing public rankings. To them it's no different than being thrust into another planet. Adapt or perish. Adapt and overcome. Adaptation requires learning. And unlearning. Innate intelligence helps, but as applied to new realms and under new and changing circumstances."

Deanna activated her iPhone230X and dictated this:

'There is one prime disruption that has been the key precept for the federation for hundreds of years."

We will never be satisfied with any healthcare disparities based on race, creed, religion, sexual orientation, socioeconomic status or planet of origin.

Bones jumped in, "Actually, I think there are two prime disruptions. That one and:"

Look at healthcare as a team sport and develop a system that is both user-friendly and delivers value and puts incentives in place to pay that team based on optimal health outcomes.

Spock chimed in, "It's hard to believe in this century that surgeons don't even have to prove they are competent. This logical disruptor came from my home planet of Vulcan:"

Use technology to prove that every surgeon can objectively prove appropriate competence and confidence to perform the requested procedure.

He continued, "and while the Vulcan part of me enjoyed the logic of how future doctors were chosen by pure science criteria alone, I found it strange that there was no necessity to be human to be a doctor."

"That's right," Bones replied. "The last thing we needed was a bunch of half human robots (no offense intended Spock) taking care of patients, so we now:"

Select and educate interplanetary physicians of the future that are empathetic, communicative and creative and let the robots and tricorders be robots and tricorders.

Troi interjected, "Well let's just inject the others into the collective mindset and get out of here. Here they go:"

Move from a hospital mindset to a consumer health mindset, that is, get care out to where people are.

Always send a believable understandable bill that clearly states what was done, what it costs and what a patient owes.

Learn from other countries (and other planets) as it relates to different modalities of healthcare. They are not "alternative" they are just 'different.'

Data is not a weapon, share it. The hospital across the street is not your enemy. Cancer and heart disease are. Fight them together.

Open source all electronic medical records so that developers can work together to bring decision support in healthcare such that healthcare providers will know at least as much about their patients as Amazon does about its prime members.

Learn from other industries and take a systems approach to employing population health in a way that actually makes the population healthier.

Captain Picard waved goodbye and said it best.

"When faced with the choice a famous wizard offered, between 'what is right and what is easy,' we have to do what is right. You need to challenge your teams to grow and change so they can adapt to any situation. You need to seize opportunities as they come so that you don't coast through your lives as physicians, nurses or anyone else in the healthcare ecosystem. Follow these lessons, and they'll take you on the next stage of exploration. For us it is where no one has gone before. For you, it's a whole universe of healthcare unencumbered by the way it used to be!"

Spock went back to the computer and said, "perhaps the humans of this century would do better taking heed from some philosophers of their own planet and century as they struggle to disrupt healthcare."

"Buckminster Fuller said, 'You never change things by fighting the existing reality. To change something, build a new model that makes the existing model obsolete.'"

Bones agreed, "If the people of that century can take our disruptors as the goal, embrace new technologies, but accentuate their humanness, then they will succeed in creating a new transformed healthcare model."

Spock continued, "A philosopher named Pat Riley once said of basketball teams, 'When a great team loses though complacency, it will constantly search for new and more intricate explanations to explain away defeat. After a while, it becomes more innovative in thinking up why they lost than thinking up how to win.'"

Captain Kirk smiled and decided he needed the last word. "You missed one Spock. He wasn't quite a philosopher but a warrior named Mike Tyson who once said 'everyone has a plan until they get punched in the mouth.' I believe the healthcare leaders in 2018 are feeling pretty beat up. Between our help and the fact that they need a new model, I have great confidence that they will stop concentrating on the past and their differences, but look toward an optimistic future. Who knows maybe this future will have them reach the Intergalactic Honor Roll for Healthcare by 2035!"

Counselor Troi smiled encouragingly, "I sense great longing among these people as it relates to healthcare. It seems as if they are ready to embrace analytics, transparency, and end to end seamless experiences to stop whining about what isn't working and join the great planets of excellent healthcare."

"Live long and prosper" was the last message we heard as the Starship and its crew faded off into space.

It worked. The USA on the planet Earth was transformed. But it didn't happen by magic. It happened through optimism, risk-taking and creativity. Most importantly, the "consumer revolution" became the "patient revolution." What started as patients wanting mobile healthcare, and patients realizing they had to pay more for their healthcare, suddenly became patients having a role as pilots of the entire system.

So whether it was the violation of the prime directive by my Star Trek heroes, or people being "mad as hell" and not taking it anymore, or healthcare leaders tiring of explaining why it wasn't working, or the "12 disruptors" from my other book, or even being "punched in the mouth," the revolution indeed happened.

Our best prediction is it happened when patients became true members of every part of the healthcare team:

- ✦ When patients were seen as core members of medical teams to help guide true personalized care "anywhere I am." When patients actually answered the question, "what would *you* do to help this person?" (as we've seen applied successfully in Type1Diabetes and Cystic Fibrosis and Alzheimer's and so many more chronic conditions).
- ✦ When patients joined scientific research teams to fight against bureaucratic barriers, advocate for funding, expand trials and join as inspirational partners with scientists (as we've seen in patient-driven research initiatives with groups like Emily's Entourage).
- ✦ When patients brought their stories and their advocacy to every medical conference and journal to reduce the flood of non-impact publication and boost the results of great conferences (as in patient participation in Stanford's MedX.)

What started with patients demanding federal funding for research (breast cancer, HIV, etc) became major partnerships with scientists. What started as patients wanting access became care at home. What started as patient engagement became a strategy for ending isolation and depression.

As the song said, the answer is "everyday people."

Now it's your turn.

We want to you use the end of *Bless This Mess* to create your version of the revolution. How do we - all of us - create the ideal future for the delivery of healthcare?

What can **you** do that may be small today, but by 2035 will result in the USA on the Planet Earth being accepted into the Intergalactic Health Council?

How do **you** become the missing link to an optimistic future in healthcare.

Use these pages to draw your ideas, your disruptions, your ideals. The sky is no limit -- you have boundless space to envision your world.

Remember … the sky's the limit (or the galaxy … or the galaxy beyond …) !

Draw your ideas here, and send them to me at sklasko.com. The best will be published in the journal *Healthcare Transformation*.

The FUTURE of Healthcare in America

A Tool for Dialogue in Your System

What do we want to SEE in 15 years?

Apply what you've learned based on life on other planets.

Draw your innovations here.

What do we have to BELIEVE for this to be possible?

BLESS this MESS

The FUTURE of Healthcare in America

A Tool for Dialogue in Your System

Starting TODAY... What can I do?

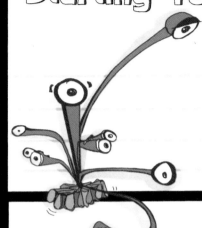

What can my team try?

Who else needs to be at the table?

BLESS this MESS

ABOUT THE AUTHOR

Stephen K. Klasko, MD, MBA, is an international advocate for transformation in American healthcare delivery and is considered by a variety of rankings to be one of the most influential thinkers in healthcare. He is president and CEO of Thomas Jefferson University and Jefferson Health, one of the fastest growing academic health entities in the United States.

Chrissie O. Bonner helps groups understand their own narrative, often by drawing conversations in real time, as a generative scribe. As a graphic facilitator, she helps organizations see their path to growth and success. She is the founder and owner of Illustrating Progress.

Michael Hoad is a former journalist who was born in Guyana and grew up in Kingston, Jamaica. He now enjoys helping institutions integrate their values into strategic positioning. He is vice president for enterprise marketing at Thomas Jefferson University and Jefferson Health.